FUNGUS POEMS

Psychedeliciella mycological var. Edinburghensis

Written and donated by the world's leading mycology scientists.

www.poemcatcher.com

Inspiration on the fly
Published by PoemCatcher Creations
Salisbury Centre
2 Salisbury Road
Edinburgh, EH16 5AB

www.poemcatcher.com

Copyright
All the poems in this book were donated with love and permission to be published. The copyright remains with the authors. This is their beautiful creativity and I am just a creative collator.

The Cover photo was taken by Greg MacVean in Hopetoun Crescent Garden, Edinburgh. http://www.hopetouncrescentgarden.org.uk/

Cover Design by Trevor at Fresh Digital
Doodles with love from Mayia Karachaliou
Psychedeliciella by Rita Verma
Prints from watercolours donated
by Peter Thwaites at www.mushroomart.co.uk

Use of this material is welcomed – providing it inspires, engages and enthrals audiences.
Each and every poem in this book is brilliant. If you disagree, send £10 with your complaint to the charity of your choice.

This is a 1st edition
It was produced in between 1st and 6th of August 2010
at IMC9 Mycology Conference, Edinburgh

"FUNGUS POEMS"
ISBN 978-0-9566018-5-8

Be proud to own it!
Excellent laboratory reading
Do expect it to sporulate.
Or propagate

Dedicated to the Art of Science

And the Science of Art

With special thanks to Nick Read
Professor of Fungal Cell Biology
University of Edinburgh

Contents

THE INVITATION 10
    Grow the art of your science (English version) 10
    Growing the art of your science (Scottish version) 12
    The invitation 15

THE RESPONSE 17
    Ode to Polyphores 17
    The Big Party 18
    A Fungus Sigh 19
    Autumn Decay 20
    The forest's floors 20
    Gary's Limericks 21
    Autumn's blues 22
    Translation of Autumn's Blues 23
    Spore 25
    Termitomyces 26
    Transformer 27
    A field haiku 28
    About Mushrooms (fungi) 28
    The Mushroom's Song (for Tatiana) 29
    Lichen on Mars 30
    Fun-Gal 31
    The Mighty Fungus 32
    Mascaria in Mind 33

- Scotland 34
- Umbrellas's 35
- RMK Lymric 1 36
- Ode to a Microbe 37
- A fuss-free fungus 39
- O Candida 40
- O cerevisiae 40
- The Mushroom Seeker 41
- If in doubt, throw it out 42
- The Tao of Commensialim 43
- The Beloved Mush-Room 44
- Puffballs 45
- We Love Fungi! 48
- Fungi's to work with 48
- Cords of White 49
- Oh to be a Fungus 50
- Of Fungi and Men 51
- French poem 52
- Stan 53
- On my plate 53

MUSICAL INTERLUDE 54
- Dear Fungus 54
- I want to be your fungi 55

THE RESPONSE CONTINUED 58

Directed Panspermia	58
The Pink Pint	59
Taxonomy Quest	60
C.a., S.c., C.n., M.g., F.c., et al.	61
Fliegenpilz fand Pfifferling	62
Fliegenpilz fand Pfifferling Translation	63
Nature's own Mycologists	64
A Spore takes flight	64
The past and present	65
Saprotrophs	65
Cheers for two old friends	66
Beowulf meets Basidiomycetes	67
The Fusaria	69
I get there!	70
Armal – Limerick	70
Reflections	71
A tale of whoa (!)	72
We had not met before	73
Hymn to timely action in mycology	74
Oo	76
There is grandeur to this view of life	77
Fungal Valentine	77
Mycorrhizae	78
Hangover (or Morning after)	78

Nomenclaturists Lament 79
Treasure Hunt 79
Ode to the Spitzenkorper 80
Rust 81
Hyphku #1 82
The poor little Mykopat lost in phylogenetic space ... 83
Playful boys 83
Oh my lovely mushroom 84
A Fungal Friend 85
Identification of Marine fungi 86
Wish of a gasteroid 87
Wondrous Fungus 88
Spore 89
The Stinkhorn 89
What's under the mycologists kilt? 90
The Joy of Dank 92
IMC9 93
Write another poem here 94
APPENDIX 95
Translation of "GROWING THE ART OF YOUR SCIENCE (SCOTTISH VERSION) 95

Dear Scientists,

Congratulations on breaking the mould!

May this book stand as testimony to the creative abilities of scientists everywhere.

Your work is so important.

You are artists in your own field. Exploring nature at macroscopic and microscopic levels. Seeking insights, connecting dots, revealing unseen truths. This takes brilliance. Your brilliance.

Your work and your lives are infused with creativity, innovation and joy. May this truth be the discovery of your science.

                Happy reading
                PoemCatcher

Laccaria proxima

THE INVITATION

Grow the art of your science (English version)

To the scientist who fears
the world of art so strange
I share with you the doubt
That e're the two shall twain

Together? be these worlds
Together? where to start?
Together? doubt our minds
Can mix a science and art

You know not of the rhythm
The meter or the form
for words to blend to meaning
for poems to write and dawn

I know not of the science
The technics of the lab
Pharmaceutics or genetics
the latest Myco fad

Yet here we are together
In Scotland's castle zone
a city proud of literature
UNESCO's writing home

So we call on Ye who visit
With science on the mind
Welcome to this city
We hope you find the time

to wander to the castle
Botanics and the seat
Where Arthur's names' interred
where people walk and greet

We hope you meet your nemesis
in challenge ye shall grow
We hope you taste the Whisky
with dram in ye to flow

We hope you learn of fungus
more than you ever dreamed
We hope you write a poem
to tell us what you've seen

So Grow your art of science
And teach us through your words
of mycologic meaning
mixed with poetic prose

By PoemCatcher

Growing the art of your science (Scottish version)

You're noo in Auld Reekie, winds snell and sae cauld
Like clansmen drawn hither in yon times of auld
When *Crann Tara* (fire crosses) their clear duty told
  Burn'd oe'r the heather;
Mycologists now – in that very same <u>mould</u>
  Come gather taegither!

You're aye very welcome to this oor braw toon
Ilka <u>*Persoon*</u> has space – there's so very <u>mush room</u>
While yon castle rock there benignly looks doon
  On folks frae aye pairts
Tak time tae yourselves, have a richt gawp aroon'
  'Fore bletherin' starts

But Mycologists fear the warld that's cried art
Believing quite clearly they should 'bide apart
So here's a wee challenge richt frae the start
  Apply rules o' the rhyme
Make wards intae verses that are canny and smart
  Like Burns in his time.

There's Rankin an' Rebus an' Rowling and mair
Mma Precious Ramotswe was also penned there
An' folks who said proudly '<u>*my college i*</u>s here:'
  Lister, Simpson and Higgs
There's even the place where they schooled Tony Blair
  And Telford's Dean brigs

Of Stevenson, Brody, an' Jekyll an' Hyde
James Clerk Maxwell, Chris Hoy and much mair beside
Who said 'I wis born here' and said it wi' pride
    Big Sean (he's The Bond)
And aither great names that for long will abide
    On shores near and beyond

Alexander G Bell – a name that should ring –
(Whose cell phone is that? turn aff the damn thing!)
And aithers whose fame and history bring
    World Heritage Status
So here's tae the place of which we all sing
    Amang cities – The Greatest

And here lies oor Dolly, the first sheep tha's a clone
(poor wee *lamb, 'ell o* a life she had known)
There's science and story in each granite stone
    And Jacobite battle
Or follow the Templars and tales of Dan Brown
    To Rosslyn Chapel

Young Conan Doyle, anaither famed son,
Get Holmes on the case – wi' Doctor Watson,
Assisting that Higg's find elusive Boson.
    Or they might go Trainspotting
Or it could be the outdoors is preferred by someone
    Go walking or yachting

Remember and _stipe_ through all those wee door
You've been _to'd, stools_ in bars are far frae the floor
Have _fun, qi'_ yerselves a dram an ain more
        _Pile'y_ous intae Tattoo
In the city that has such a Volcanic core
        And nae say it'_s poor_

Ye can see that this toon mixes Science wi Art
It gangs way back in time 'fore Noah built his Ark
So all that I ask you – come play a wee part,
        Get something writ doon
And let's get it done noo, afore you depart
        Back tae your ain toon

By Simon Maclaren

See appendix for a translation

The invitation

Mushroom, mushroom poem
Mushroom, mushroom, mushroom poem
Write a mushroom poem

By PoemCatcher

Boletus erythropus

THE RESPONSE

Ode to Polyphores

Poem Title: Ode to Polypores

Polypore, polypore
highly holey xylovore
leathery, crusty, from spores all dusty,
long-lived hymenophore.
You I do adore!

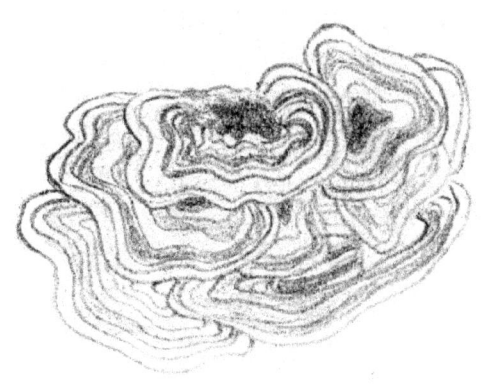

By Veera Norris

The Big Party

Seventeen hundred and sixty mycologists strong
Travelled to Edinburgh to join in the throng
Of fungi biologists of every creed
Invited me to the Big Party, thrown by Nick Read.

The EICC reverberated to the big mycota bash
All day long at the lectures – then out on the lash
The data, the bar charts, the graphs – oh what a feast
Of filamentous hyphae and the budding of yeast.

The audience sat gripped, stopped in their seats
Enthralled by the movies of ascomycetes,
A short break perhaps for a coffee or tea
But lets get back to it quick – for more GFP.

A festival of fungi – but just for a week
Then sadly its back to the grind (so to speak)
But we'll all go back happy, and stronger and wiser
That IMC9 party really inspired us.

By Neil Gow

A Fungus Sigh

Poem title: A fungus sigh
author: Libor Mrnka

It's not that easy being brown
to be entraped 'tween dusk and dawn
to strive with green in every way
under its shade I weap, I pray

To flee from light to dark of soil
a lonely twine of twisted coil
face endless changes in search for source
it's truly cursed by God the course

Yet to the midst of my face frown
a beam of hope broke through the dome
don't fear to be dependent, foriend
no gloomy, it's by far more splendent end

For just as Men for gold, the light
for what jetr you crave they bleed in fight
from iron, bronze are two sides of surface
~~thus~~ adopt advice to escape the furnace ~~of thins~~

Hold branched
~~Take the firm~~ hand of your green foe
stand together, form an army row
make the way as two edge sword light heated
I swear you'll never be defeated!

Autumn Decay

Under the brambles and the berries
A sweet scent of decay
Rises as the secret kingdom
Stirs and the fungi return to feast
Before the cold tug of frost.

By Geoff Robson

The forest's floors

Creeping, seeping, spitting spores
Fungi link the forest's floors
Living fountains, shifting webs,
Feed the living, eat the dead.

By Luke Heaton

Gary's Limericks

He came to the conference by bus
His nickname at home was fun-Gus
Originally shy
Angus was now a fun guy
But it took years to be humourous

IMC9 was held in Scotland
The delegates formed a great band
Themes from culturing to disease
All enjoyed with great ease
If only such joy could be canned.

IMC9 speakers came from far away
To present their work in a way
To inspire delegates
Not about ciliates
Only fungi can make us all stay

IMC9 delegates came from afar
Some fused and moved to the bar
Looking filamentous
Just like a fungus
They swarmed as if on agar

By Gary Jones

Autumn's blues

Als de herfst
niet meer te ontkennen valt
komen de paddestoelen
die ik met een dof gevoel
begroet.

Misschien zijn ze als een
kleine troost bedoeld,
voor het verlies van warme dagen,

maar soms
als het weer te mooi is
schijnt het licht zo sterk
op een gekleurde hoed,
alsof de zomer zich nog
eens bedenkt.

En als de beslissing –weer-
gevallen is,
nemen jongens wraak
en geven elke vliegenzwam
een rotschop....

de mooiste het eerst

Translation of Autumn's Blues

When autumn
cannot be denied anymore,
the mushrooms arrive,
which I welcome with an
ambivalent feeling

Maybe they are meant
as a small consolation,
for the loss of warmer days,

sometimes
when the weather is too beautiful
the lights shines
so strongly on a
coloured hat
as if summer
reconsiders once again.

When the decision
Has been made
 –again-,
Boys take revenge
and kick every Amanitus (??)
they can find....

the prettiest are first.

By Jan Dijksterhuis

Hygrophoropsis aurantiaca

Spore

Spore,

My eye
can't spot
what
you tiny ball
called spore
can do

Green seen
On my bread
Lead me
To your existence
You who
Poisons it.

Ball
So small
And green
Unseen by me
Fly by wind
To gain in power

Shaping apical tips
To tube the wolrd
Your tubes grab food
And branch
To cover land
To big to oversee

My eye
Can spot
What
You tiny ball
Has done.

By Henk Huinink

Termitomyces

        Emerging from the funguscomb
        Just like a child from a mother's womb
        You are loved by termites
        And are devoured by mankind
      Your size is variable,
        Thus making you desirable
O, Termitomyces, Whyast thou so rare?
        To culture you is a nightmare
        Because your growth is a wonder
   Thus making mycologists to ponder

By Ginay Mathew

Transformer

Poem Title: Transformer

~~Forge~~
Damp, warm, weather, storm,
Growth, ~~understood~~, earth, decamp,
Transform, mutate, ~~static~~
And Speculate

By Russell Paterson

Creativity emerges.
Scratching its way to clarity
Given the invitation

This is a favourite poem,
For all the possibility it shows.

A field haiku

Mushroom, mushroom, mush
Room, mushroom, mushroom, mushroom
Mushroom, where are you?

By PoemCatcher

About Mushrooms (fungi)

I'm hate & afraid of them
And I like them
Very much

By Komaletolinova Elvika

The Mushroom's Song (for Tatiana)

Awaken from your sleep
Dream while you're awake
This is the nectar
We want you to take

There is no time
There is no space
There is no way
To measure this place

There is no new
There is no old
There's never been
Any story to be told!

To succeed is to travel
From one to another
Look within yourself
And you will discover

There is no you
There is no me
Only the tendency
Of energy to be

We become one
There's nothing to fear
Let yourself go
So we can be near

Feel the heartbeat
Of the buds on the tree
The stone and the earth
Your lungs when you
breathe

In air we are everywhere!
The pulse is the rhythm
Mycelium to spore
Who travelled from stardust
To open our door.

By Michael Finnie

Lichen on Mars

The way that I survive
It is by doing nothing

Not what you would expect
From one living at extremes.

By Hans Blom and Louise Lindblom

Fun-Gal

They can be plant killers
And some will help plants
Some can catch nematodes
And some are formed by ants
They're found in space stations
And three week old tea.
Some eat jet fuel,
Others live in eth sea.
They live off of leaves
And the human's dead skin.
There is no place
They are not found in
Without them on Earth
Our existence would pall
So I'll sing their praises
I'm the ultimate fun-gal.

By Melissa Day

The Mighty Fungus

Sheltered in the forest shadows

        Growing hyphae all around us

Fruiting in the blooming meadows

        Unrevealed and mighty fungus!

From petri dish to metagenomes

        Mycologists spread their wings

Beneficial?? Full of venoms??

        ...amazing what fungus brings!

By Jiri Jlrout

Mascaria in Mind

To find the sacred mushroom
on the trail,

The 'shroom of shaman,
 priest and fairy tale

To seek and spot the red
topped Amanita

And then, oh joy
To pick and eat her

By Ted Tuddenham

Scotland

If you are at a beautiful place
& you cannot believe that you
Are alone there, you have two possibilities:

a) Go one further meter
    and everything gets horrible

b) You are in Scotland &
    Share everything only with
    Rabbits and fungi

By Nicole Remme

Umbrellas's

The Sun's ray fell on the earth...
    Spreading its radiance of wealth...
Birds chirped, welcoming the air of warm breathe...
    The green grasses dripped with the juicy nights dew...
Heaven's Rainbow had sprayed its lines,
    Breaking the boughs with varied hues
Beneath the trees tiny little sprouts are in view
    Tiny little – like mini umbrellas of shade
Some white and some with tiny dots made
    Oh! What a wonder with no green leaves
Yet they out do all the heavenly flowers
    They are tiny fairy umbrella-shades of heaven
They blossomed at height for elf's to dance under them...
    Rejoicing the rhythm of life...

By Dr. S. Swatha

RMK Lymric 1

I arrived on a plane in Scotland
The weather took a turn for the rotten
I came for the fungus
And stayed for the haggis
The whisky and ale made me sodden

By Ryan Kepler

Ode to a Microbe

Our microbe is so very small
We cannot make him out at all
But many sanguine savants hope
To see him through the microscope

Upon his walls a pattern stands
Composed of many disparate bands
Of interwoven beta glucans.

So, push and pull and probe and poke
Our nanoscale research bespoke,
Send image to the digitizer
And chuck in data from ELISA

Atomic Force Microscopy
Gives imagery for all to see!
A truly Systems Strategy
...and fundable by BBSRC.

Architecture, much conjecture
Surely time for end of lecture?

By Sarah Gurr
(with apologies to Hilaire Belloc and John Taylor)

Boletus edulis

A fuss-free fungus

(OR: Why Sordaria macrospora is a good model organism to study fruiting body development)

Fruiting bodies (we all agree)
Are wonders of multi cellularity
But how fungi manage to regulate
The production of structures to propagate
Is something we have no clue about,
And research is needed to figure it out.

But which fungi to study? The choice is quite hard.
What about mushrooms for a start?
But even though some of them are quite tasty,
Their choice as a model might be somewhat hasty:
Weeks, even months their growth may be,
And many aren't happy without a tree.
What we need is a fungus that makes fruit bodies quick
And needs no partner to do the trick.
A fuss-free fungus is what we need,
And there's one example that can hardly be beat:

Fruiting bodies in less than a week –
Sordaria is just what we seek!
And it doesn't need a special diet,
A bit of corn meal is all that's required.
Just add some barley malt (but no hops)
And the fun with this fungus
never stops!

By Minou Nowrousian

O Candida

As you kissed goodbye,
    Your birth scarred your mother

Now you are expectant too,
    Ready to bud

Wanting to transform and invade

With sinister purpose
    You patiently wait

By Gavin Sherlock

O cerevisiae

More elegant than your
    Distant skizo cousin,

You're a beautiful model
    Who knows how to make beer

What could be more perfect?

By Gavin Sherlock

The Mushroom Seeker

In search of the fungus I go
Into the woods, searching, sleuthing
Looking to find the place they grow

Ahh,
under the oaks
I see
A giant one pokes,
disturbing the earth
The tell-tale sign
of the wonderful treat
That I seek.

By Diane O. Inglis

If in doubt, throw it out

Amanitas can be red
Lactarii can be too
Agaricus Augustus is sweet,
If you ask "Can I eat?"
I won't have a clue

By Jackie D

The Tao of Commensialim

I wish, I were me fungi waiting back in my lab
I'd be caved for night and day, and petted step by step
I wouldn't envy me cousins – living wide astray
There's nothing like being plated out, after a hard works day.
Master and I make papers – which are even read!
Buts sometimes I do fear, he morphs into a labrat
Sometimes if we play to hard, and things get just a mess,
Then we discover funny things and fly off to compress.
We are one shadow – we grow together older,
He disinfects his hands a lot – but I! lurk on his shoulder!
I pray for him, my heart heats for him warmer
Killing us sometimes in thousands – means certainly had karma
So as conclusion – who knows when and why –
I finally might be reborn – as my own lab fungi!

By Matt Leonhard

The Beloved Mush-Room

Where would I like to stay all day?
On the mush-room
In golden-green gloom

On it's lush mossy cushions
Under the pine canopy
Sparkling drops of dew
Bird songs – a few
A gentle creek flowing
Rowan berries glowing
Like gems
That's where I'd like to stay –all day.

By Parnassia

Puffballs

As kids
    We ran through the woods
    Searching you out
When found
We kicked
    The living daylights
    Out of you
Sending plumes of smoke
    Into the air
And coating the cuffs
    Of our trouser legs
    With sticky, green spores
Which drove our mothers
    Up the wall.

By Gavin Sherlock

Coprinus comatus

We Love Fungi!

We came to Edinburgh, leaving our house
Taking a couple of planes and the PC mouse
To the congress center walked,
And about fungi talked
Because we love fungi-things
But also human beings!

By Sandra & Sara

Fungi's to work with

I had some sequencing to do
It made me feel a little blue
There are so many companies
From which to choose
I wanted the best one with whom
I wouldn't lose
I had a thought & in a sec
I decided to use G.A.T.C. Biotech!
With the orange company I assume
They could sequence my mushroom
The service they gave was the best
I'd recommend them to the rest

By James Archibald & Catherine Scerri

Cords of White

Cords of white beneath the leaves

No eyes to see or feet to walk

No ears to hear of tongues to talk

Just endless arms in velvet sleeves

By Brett Arendz and Robert Blanchette

Oh to be a Fungus

I'd love to be a fungus
A teeny weeny mould
To tip elongate and sporulate
And never grow old

By David Guest

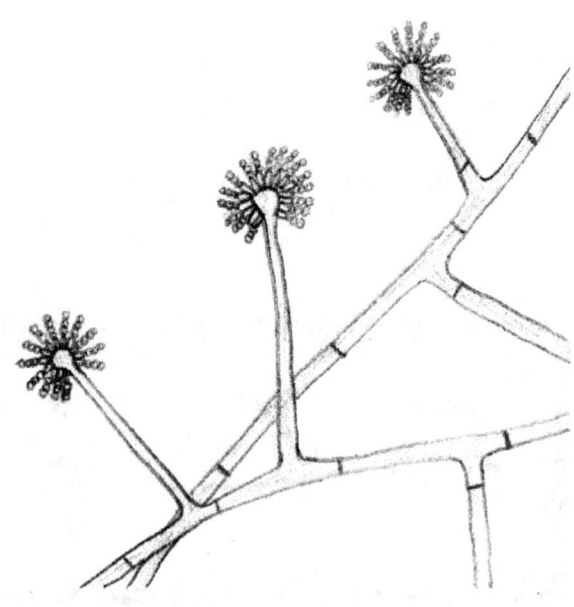

Of Fungi and Men

Mushrooms and ascii
Hyphae and spores
The art of nature
Everyone adores

Yeasts go for brewing
Don't' like the cold
What would the world do
Without the mould?

Next came the lichens
Last but not least
And mycorrhizae
A symbiotic feast

'n dear Aspergillus
My favourite pet
Can't live without him
Nor you , I bet!

By Mayia Karachaliou

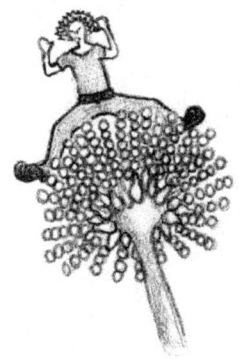

French poem

Il n'est jusqu'au climat
qui ne soit favorable aux champignons
dans cette belle cité dont on aima
le temps d'une semaine la fongique opinion

By Sclosse

Stan

There once was a fungus names Stan
A man fried him up in a pan
    Then, in one quick bite
    Poor Stan had a fright
And "yum, yum" said the well fed man.

By K. Plett

On my plate

Stuffed, sautéed, baked or fried
In a sauce or in a pie
Portobello, cremini
Or a truffle to try

Mushrooms in all their forms
My appetite can sate
So long as they are found
on my dinner plate

By K. Plett

MUSICAL INTERLUDE

Dear Fungus

Dear Fungus

Dear Fungus
I don't think I love you anymore
You used to look so lovely
Now you're rotting up my floor
Dear Fungus
I don't love you anymore

It used to be you and me
Some drawings and light microscopy
that's all we needed to get by
Now it's molecules, genomes and cells
and death by DNA phylogeny
Dear Fungus
I don't love you anymore

Now the Fungus sings:—
Dear Professor
I don't think I love you anymore
Because when you get researching
You're probing places I've never probed before
Oh Professor, I don't love you anymore

Why don't we get together
and get a grip on our human-fungal relationship
We could almost be symbiotic if we try
We could grow along together
and live happily for ever

Dear Fungus I think I love you after all

Oh and dear Professor, when researching
don't breathe my spores

Geoff Gadd © 2010

I want to be your fungi

I want to be your fungi
I wish I was a mushroom
Growing in your park
Or you could put me on horse manure
and squeeze me in the dark
I wish I had ballistospores
I'd aim them at your heart
But I've got defective pheromones
They're keeping us apart

I wish I was a bracket
and you a rotting tree
I'd degrade your lignocellulose
and let your carbon free
I could be your cytoskeleton
that would hug your organelles
I could even be a stinkhorn
and you could be my smell

Ch: I wish that I had hyphae
that would colonize your world
I want to be your fungi
but will you be my fungirl

x2
instr

Could we try anastomosis
and let our nuclei fuse
Differentiate a fruiting body
what have you got to lose?
We'd colonize more substrate
Producing lots and lots of spores
I'm willing to give up everything
So that my genome fits with yours

Ch

Your basic hyphal growth unit
is as cute as cute can be
I could be a partial differential equation
and model your symmetry
I wish I was an endophyte
and you a leafy tree
But it would have to be an evergreen
so that you would not lose me
ch
last 2 lines twice

Geoff Gadd © 2010

See Geoff's open Myc performance of "Dear Fungus" on http://www.youtube.com/user/Poemcatcher

Coprinus picaceus

THE RESPONSE CONTINUED

Directed Panspermia

I have heard the humble mushroom spore
can travel to earth via meteor
able to survive both fire and frost
boundaries of space and time are crossed

Leading McKenna to hypothesize
on reasons why we are so wise
the curiosity of some ancient monkey
who stooped to nibble a cosmic fungi

An interaction with mushroom kind
which stretched his noggin, blew his mind
higher intelligence had landed.
human consciousness expanded.

By Kate Masters. Edinburgh.

***Panspermia** is the hypothesis that life exists throughout the Universe, distributed by meteoroids, asteroids, and planetoids.*
***Directed Panspermia** suggests that the seeds of life may have been purposely spread by an advanced extraterrestrial civilization (Wikipedia).*

The Pink Pint

> **Poem Title:** The Pink Pint
>
> In the beginning there was the SPORE
> dancing in the air.
> In front of a bottle of beer
> fungi suddenly appeared.
> Looking deep in the bottle
> fungi smiled and settled.
> Fermentation was very good
> and words raised and then stood.
> The beer disappeared as well as the bean
> We started to be like a pink pig and
> ... TO THINK BIG!!!

By Hyco-Genesis 1:1

Taxonomy Quest

Poem Title: Taxonomy quest

Laetiporus sulphureus
　Physisporinus vitreus
Cryptomycocolacales
　Biaportales
Thelephoraceae
　Amanitaceae
Schizopora Phytophthora
Abas Raphas Epicnaphus
　unsolved chytridiales

Please write your poem here and drop it in the PoemCatcher Net.

By Ewald Langer

C.a., S.c., C.n., M.g., F.c., et al

Poem Title: C.a., S.c., C.n., M.g., F.c. et al.

They can be road, they can be long.
They arrived first, and will leave last.
Their legions have explored continents.
Invisible, delebratly exposed,
the fruit of their love gives us joy or fear.
They are used and sometimes abused,
but their loyalty will stay to the End.
They are with us from start to the finish,
and will always be looking after us.

By Cottier Fabien

Fliegenpilz fand Pfifferling

Poem Title: Fliegenpilz fand Pfifferling

Ein Fliegenpilz stand reglos da
und suchte nach einer anderen Amanita muscaria.
Aber leider fand er nur einen Pfifferling,
der an einer Buche hing.
Der Pfifferling sagte: "Es tut mir Leid,
aber das geht wirklich zu weit.
Ich hänge jetzt schon Stundenlang
an dieser blöden Buche dran.
Kopfüber zu hängen und in der Sonne zu versängen
ist nicht erträglich und sehr schädlich.
Darum Leute sag ich euch: Ihr tut euch
gut daran nicht zu tun was der Pfifferling
getan.

Please write your poem here and drop it in the PoemCatcher Net.

By Martha Langer
(12 years old)

Fliegenpilz fand Pfifferling Translation

FLY AGARIC MEETS CANTHARELL

A fly agaric standing there,
Looking for another here,
But there was only a Cantharell
Hanging on a beech tree well,
Cantharell says:" I am sorry but I will worry",
I am hanging now for hours
On this stupid beech tower,
Hanging top down in the sun
I cannot resist, it is no fun.
So fellows what I have to say:
Stay away from the Cantharells way

Translated by Dad!
Ewald Langer

Nature's own Mycologists

Maybe the closest kin to dedicated mycologists
Are the leaf cutter ants specialists
These ants collect, exploit and care for fungi
Worthy a special feature article in "I AM a Fun- guy" **

** The new IMH Journal is called "IMA Fungus"

By Lene Lange

A Spore takes flight

Poised on its little stalk
    The spore, of course, cannot walk
It awaits that magic time
    When Buller's drop becomes so big
It spreads out and has to climb
    Across the spore and with a jerk
Carries the spore just far enough
    To fall free down the trough
Of gills or pores to the great outside
    So taking off for the far and wide.

By M.O. Moss

The past and present

A fungus formerly called bolete
Felt almighty obsolete
At a congress called IMC nine
Where his sequences were all fine
But fungal names for clades were
                no longer felt as a need.

By Marijke Nauta

Saprotrophs

Whenever things are right for me
That is where I'm sure to be
You think there's nothing in the air?
A million mini-me's - one there!
I choose the place I'm going to grow
The food I need, the food I know –
Proteins, glucans, wood or fats
I wreathe in mushrooms, net and mats
A food, whatever its description.
Analyse my RNA –
You'll see my menu for today
What when its all gone, you say?
Then I'll make spores, and fly away!

By Sarah Watkinson

Cheers for two old friends

No brew would we have today
Without our yeasty friends
Saccharomyces cerevisine
And
S.carlsbergensis
So let us toast to them,
And for the time,
Move any sad thoughts
Away

By John Arendz

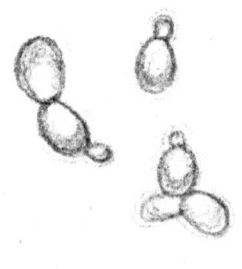

Beowulf meets Basidiomycetes

With apologies to Seamus Heaney – Translator of the ancient English poem 'Beowulf'

Under the dark fir forest
A network nurses seedlings
A network of anastomosing threads
Basidiomycetes – bringing goods for growth

Saplings spring up, turn into towering trees
Photosynthesis feed the wood-wide-web
Of fungi, keeping the forest fertile.

By Sarah Watkinson

Clathrus archeri

The Fusaria

Poem Title: The Fusaria

T
h
e
are Fungi, which are
ubiquitious
Saprobes and pathogens.
Acrobatic leapers,
Rain or air dispersed.
In soil, in plants, in us
Able foragers of the microworld.

By Kim Hammond-Kosuck

I get there!

Eat and grow, eat and grow!
What is it all about this flow?
It is to the tip the way to go!
Oh no!, but my calcium is too low!

By Sandra Mathioni

Armal – Limerick

There once was a fungus among us
Whose basidiocarp was humungous
Transgenics, it seemed,
had rendered it mean,
and it stretched out its arm and it flung us!

By Kelly Craven (et al)

Reflections

Fungi, fungi everywhere

But not a one to eat.

By MC

REFLECTIONS

A tale of whoa (!)

Woe to the bats with white noses in caves
Woe to the rice with no resistance to save
And amphibians consumed by chytrids it's quite clear,
We may not see too many frogs next year,
And what about candida,
Which in infection plays a great role?
Taken together, it all seems out of control!
But don't be alarmed,
this is not meant to depress,
1760 IMC scientists
Will clean up the mess

By Nicole Donofrio

We had not met before

We spent a year ,or two
In the dark,
While blind fish mated –
Sex and opposing,
But no offspring was produced.
Was it avoidance? A deletion?
A wall that splintered our
Hyphae?
The exocyst? The kinograph?
Shupees nidula, Aperi nidu?
Magnaporthe oryzae or grisea?
Ask TUNEL, FRAP, FLIP,
Array, ChIP-chip:
We just haven't figured out
Just what it means to be.

Anonymous

Hymn to timely action in mycology

By global efforts mycologist found 100,000 fungi
Since back of time
In the last decade we got 1,200 more
Online

Now we only have 1,4 mill more
To do
Shall it take more than thousand years
Partou?

The world may have changed
Beyond match
Will someone then start over
All from scratch?

We better decide immediately
To speed up significantly.

By Lene Lange

Cantharellus cibarius

Oo

Oo, a fungus look-alike
That is a creature that I like

It is fond of salmon, potatoes and oak
But this intimacy often leads to a stroke
Oo's new friend suffers and cries
And it won't last long before it dies

That is why Oo is a friend and foe
Its dynamic nature often elicits "Oh-oh!"

By Francine Glover

Written on a napkin

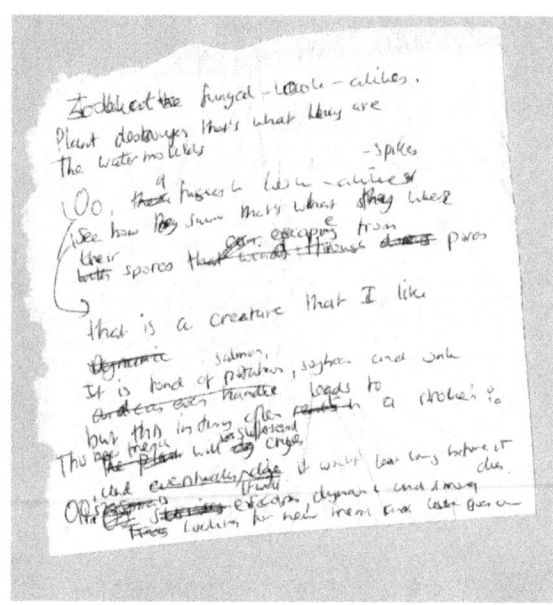

There is grandeur to this view of life

I love to drift my boat
Amongst the speciation islands
Who knows where it goes
Or from whence it flows
I drift and I wonder
At the splendor
Of the grandeur

By Bernard Slippers

Fungal Valentine

Fusarium is red
Penecilium is blue
Fungi are fab
And mycologist are too

By Jo Taylor

Mycorrhizae

Fungi's Wood-Wide-Web
Linking all those hidden roots
Their truffles call me

By John Taylor

Hangover (or Morning after)

Barley, yeast and fire;
Fermentation's amber bead
Burn's Night memories

By John Taylor

*Burns mentions amber bead in his poem
"To a Haggis". It refers to whisky*

Nomenclaturists Lament

Oh boy, oh boy, oh boy, oh boy!!!
Why do you , o Code, bring me such joy?
I fail to understand! I need an analyst!
For could it be...I'm just a poor old masochist?

By J.C. Coetzee

*After discussions to change the Int. Code
for Botanical Nomenclature*

Treasure Hunt

In a dark wood
Where no eyes can see
You could think that nothing's there
But no, don't just leave

Turn around a piece of bark
That lies on the withered leaves
Dig a bit with your Swiss knife
To expose the underneath
There, amongst the bugs and roots
A truffle waits for you and fruits.

By Heidi Tamm

Ode to the Spitzenkorper

Spitzenkorper, at the magnificent dome,
You have intrigued us for so long!
How could have we imagined what were you there for,
When nobody knew what you were made of.

But everything started slowly to get clearer.
Your sensitive nature required
Gentle manipulation and the appropriate method,
To see you at the tips of growing hyphae

Vesicles and cytoskeleton elements,
In such a delicate balance
To control your appearance and behavior
In such a coordinated way
To create that sophisticated form

Microscopists, mycologists and molecular cell biologists,
We are all fascinated by the mysterious mode,
A corpuscle with such sexy name,
Is born, perpetuates and remains at the dome.

After you went from grey to color
After you socialized with the polarisome.
Your humble presence became prominent,
And we, full of pride,
Celebrate the joyful event.

By Meritxell Riquelme

Rust

If you want to find some rust,
One thing that will be a must,
Is to turn around the leaves,
And you'll see that there it is:

Orange spots on leaves so green,
That's how it has always been,
Research it, you won't regret,
There's a sequence now in internet!

Anonymous

Hyphku #1

Poem Title: hyphku #1

threads across the wall —
the chemistry unravels:
the un-knit of mould

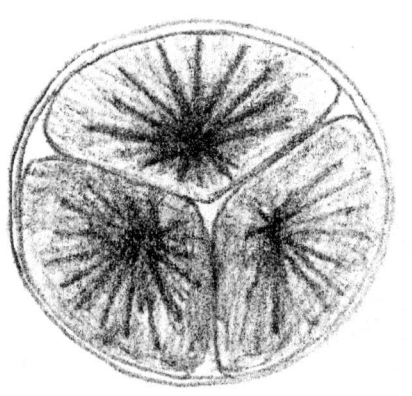

By Richard Summerbell

The poor little Mykopat lost in phylogenetic space

A poor little myckopat from Sweden
Travelled far to IMC9 to see, then,
- A fungus conquer space
- and a novel clade displace'd?

She was not quite sure whether that was forbi'den

By Anna Rosling and Karina Clemmensen

Playful boys

Oh playful boys of the forested groves
With caps all alone or fruiting in droves
All colours from browns to bright blues and red
Beneath are the searching, invisible threads
Connecting all plant in the life and death plan
Continue, caring not for endeavours of man

By Elizabeth Sheedy

Oh my lovely mushroom

Oh my lovely mushroom
How I long to see you soon
Over the Autumn season
And give my life a reason

To pick you and your brothers
Is better than any long Summers

By Neil Gow

A Fungal Friend

The world struck gold,
With Fleming mould,
"An accident", he said,
But where'd we be,
Without the key
To all our daily bread?

A fungal mate,
Makes cheddar great
But, North, south, East or West
No gracious host,
Could make his toast,
Without this precious beast.

Chantrelle is fine,
And when you dine,
Shitake's quite a feast
But, lad or lass,
Come raise your glass,
To thank the noble yeast.

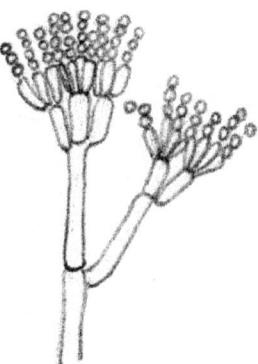

By Derek Wilkinson

Identification of Marine fungi

- Cry for help

The oceans' wide –
The fungi rare?
But you will find them everywhere;
The questions easily arise –
What are the new friends names
That the researcher claims?
So, searching all the unexplained
The ITS search bored
'Why' – this was the researchers cry –
There is no naming tag
At all the isolates in rack.

By Antje Labes

Wish of a gasteroid

Don't play football

With your puffball!

By Jan Borovicka

Wondrous Fungus

Wondrous fungus
We have admired you all week
Your capacity to sustain
Plant life and decay
Your beauty in the autumn wood
Your mystical life
Your numbers and names yet unknown
Wondrous fungus

By Rebekka Artz

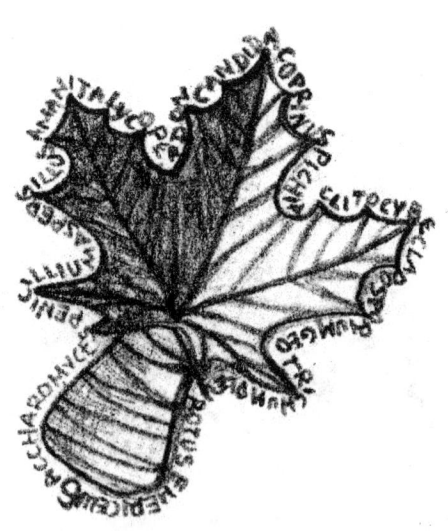

Spore

I am a basidiomycete

My spores I shoot non-stop.

"How do I do it?" you might ask...

Simple! With my Buller's drop

By Mark Fischer

The Stinkhorn

(tune of Twinkle Twinkle Little Star)

Engorged erupting biomass
Thrusting up above the grass
Reaching up towards the skies
The tip attracting hungry flies
Proud enough to shock the nurse
The impudent phallus, here in verse

By Geoff Gadd

What's under the mycologists kilt?

Did you ever go a'wandering
In Scotland's land of green
Looking for rare fungi
To eat or make you dream?

In the middle of eth summer
When the Scottish midge is here
And its ferocious bite makes promise
Right through your clothes to tear.

And you bend doon to look closer
At the fungus that yee've foound
Perhaps the first to find it
Scientific hallowed ground

And as your eyes, they focus
On the precious myco-spore
Appear a swarm of midges
Like you've never seen before

And swallowed in a breath
They fly right in your eyes
And throughout all the coughing
You miss the big surprise.

For moments after clearing
This prolific pack
You realized it was a decoy
And they really did attack

Below the kilt the army
Has swarmed around yer kit
And the agony is growing
Its impossible to sit

For the stinging of the nettle
Or the buzzing of the bee
Compares to fluffy pillows
When the midge has bit yer tree

I tell you, "Only once in life"
A man should know such pain
As the midges, mighty, mouthful
Bit on his precious gain

And you ask of me, quite honestly
If I don beneath my kilt?...
And I tell you "ONLY ONCE IN LIFE"
I'm protected to the hilt.

By PoemCatcher

The Joy of Dank

Of all that slithers
Oozes, melts and pools
Our friends & those less friendly
I would give up all of my
petty misgivings
if I could just give fungus a hug

Author unknown

IMC9

Now after IMC9
I will be fungal free

I thought I was going mad
When organization went bad

But we can truly see
The fungal majesty
Of the Kingdom!
(that are we)

By Nick Read

Write another poem here

APPENDIX

Translation of "GROWING THE ART OF YOUR SCIENCE (SCOTTISH VERSION)

You are now in Old Windy (Edinburgh) with its winds quick and cold
Like clansmen in olden times
When the Fiery Cross (Crann Tara is the Gaelic) was sent across the heather
showed them their duty (to rally to the clan Chief)
you've been drawn here
You are all very welcome to our beautiful town
Each Persoon (famous Mycologist?) has enough room
The castle rock looks benignly down on people from all parts of the world.
Take time to yourselves, have a good look around
Before the chattering/conference starts

But Mycologist fear the world that is called art
Believing that they should stay apart
So here is a small challenge right from the start
Apply the rules of Rhyme
And make words in to verses that are clever and clever
Like our national bard, Robert Burns, did in his time here

We have Ian Rankin who wrote the Rebus detective novels, JK Rowling (Harry Potter) and more
The stories of The Proprietor of the No 1 Lady's Detective Agency of Botswana – were written by Alexander McCall Smith here
Many famous people contributed to the life at Edinburgh University, Joseph Lister (Antiseptics), James Young Simpson (Anaesthesia) and Peter Higgs (Higgs Boson theory) (to name but a few)
Tony Blair was schooled at Fettes College
And the Dean Bridge was built by the famous Scottish Engineer, Thomas Telford

Robert Louis Stevenson was the author of Treasure Island and The Strange Case of Dr Jekyll and Mr Hyde possibly based on the story of Edinburgh's Deacon Brody (there is a pub named after him on the Royal Mile)
James Clark Maxwell is in the line Newton, Faraday, Einstein (the latter had pictures of the other three on his wall) for his work on electromagnetic theory, synthesizing all previously unrelated observations, experiments and equations of electricity, magnetism and even optics into a consistent theory
Sir Chris Hoy has four Olympic gold medals for cycling including three at Beijing
Sean Connery is the only true James Bond
These and more were all born hereabouts

Sir Arthur Conan Doyle (the author of Sherlock Holmes) lived his earliest years in Picardy place.
Peter Higgs, theoretical physicist, emeritus professor at the University of Edinburgh
Trainspotting was the 1996 Danny Boyle film about heroin addicts in this town
I was not able to squeeze in a reference to Eric Liddle (lived and studied here) and Chariots of Fire – also filmed here.

According to Wikipedia there are a couple of references to the parts of Mushrooms in this verse. A wee bit of poetic licence or Magic required here? (that was the only reference to Magic and Mushrooms in this whole poem)
For chariots of fire

By Simon Maclaren

www.ingramcontent.com/pod-product-compliance
Lightning Source LLC
Chambersburg PA
CBHW071831290426
44109CB00017B/1798